MONEY

Grades 1-2

REM 536A

A TEACHING RESOURCE FROM

REMEDIA PUBLICATIONS

www.Remedia.com

REMEDIA PUBLICATIONS, INC.
SCOTTSDALE, AZ

BLACKLINE MASTERS

AF797242

RESEARCH-BASED ACTIVITIES
Supports State & National Standards

This product utilizes innovative strategies and proven methods to improve student learning. The product is based upon reliable research and effective practices that have been replicated in classrooms across the United States. Information regarding the Common Core State Standards this product meets is available at www.rempub.com/standards

INTRODUCTION

MONEY features primary tasks designed to provide practice and repetition for young learners. The activities include identifying coins, learning the value of coins, counting coins, matching coins to prices, different ways of showing an amount, and simple addition of money. The book begins with a simple introduction to coins, and then graduates to practical application activities and beginning word problems.

The exercises are suitable for students in grades 1-2 who would benefit from additional practice or enrichment.

CONTENTS

Name ______________________

Name The Coins

penny
= 1¢

nickel
= 5¢

dime
= 10¢

quarter
= 25¢

half-dollar
= 50¢

Write the names of the coins.

1.  ______________

2.  ______________

3. 

____

4. 

____

5.  ______________

6. 

____

7.  ______________

8.  ______________

9.  ______________

10. 

____

Name ______________________________

Name The Coins

Complete the sentences with the correct coin names.

1. 	Here is one ____________________.
2. 	Here are five ____________________.
3. 	Here are two ____________________.
4. 	Here are two ____________________ and one ____________________.
5. 	Here is one ____________________.
6. 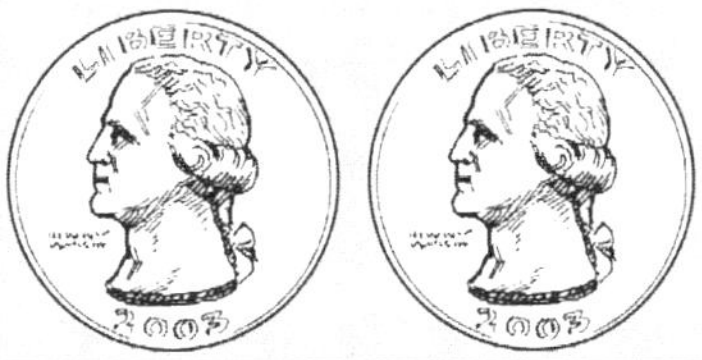	Here are two ____________________.
7. 	Here is one ____________________.
8. 	Here are two ____________________ and one ____________________.

Name ______________________________

Heads & Tails

Draw a line to match the coin Heads to the coin Tails. Then write the name of the coin on the line.

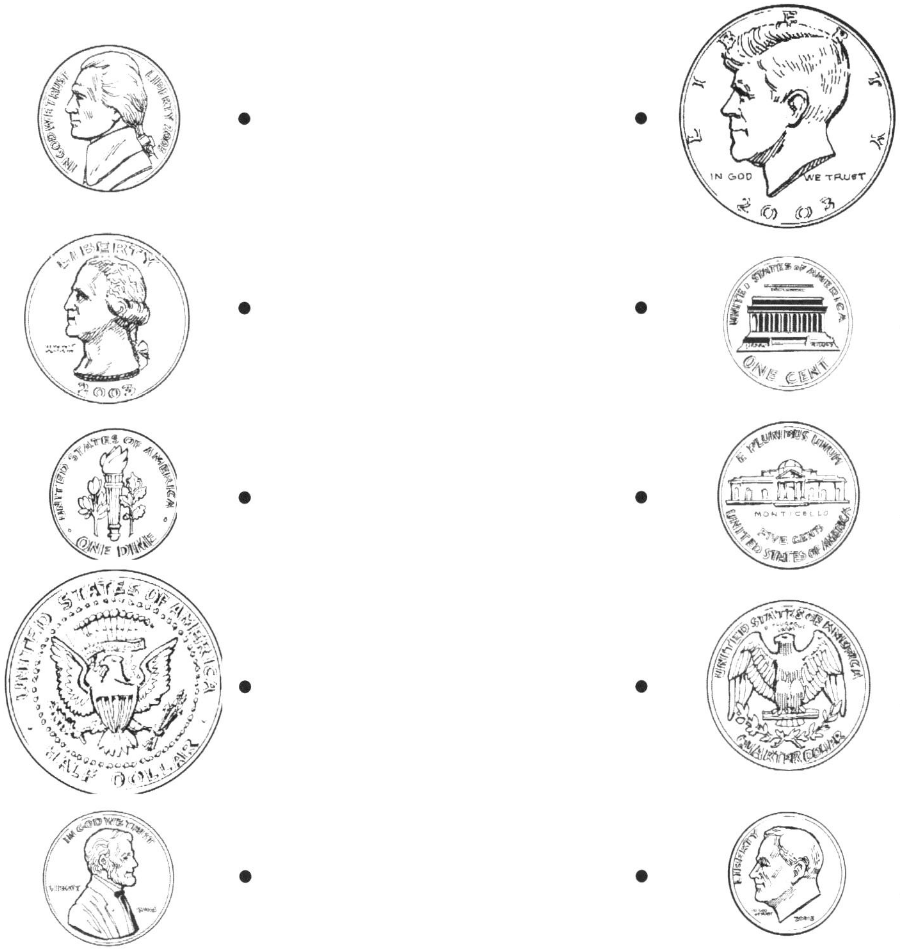

1. ______________________

2. ______________________

3. ______________________

4. ______________________

5. ______________________

Name ________________________________

Pennies & Nickels

penny = 1¢

nickel = 5¢

= 5¢

= 5¢

Count the coin values. Write the total amount on each line.

1.  = __________¢

2.  = __________¢

3.  = __________¢

4.  = __________¢

Name ______________________________

Tell How Much

Count the coin values in each box.
Then write the total amount on the line.

= ___9___ ¢

1. 

= _______ ¢

2. 

= _______ ¢

3. 

= _______ ¢

4. 

= _______ ¢

5. 

= _______ ¢

6. 

= _______ ¢

7. 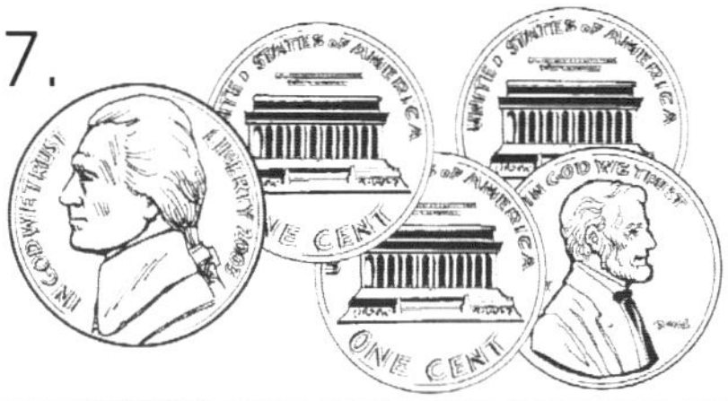

= _______ ¢

8. 

= _______ ¢

9. 

= _______ ¢

10. 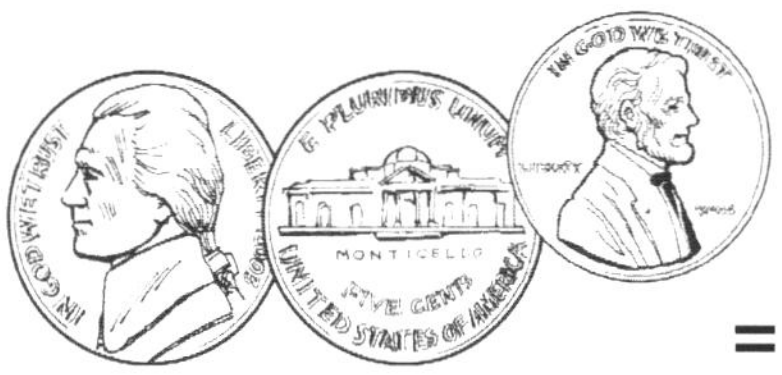

= _______ ¢

Name ______________________________

Counting By Fives

Count by fives to tell how much each group of nickels is worth. The first one is started for you.

1. __5__ ¢ __10__ ¢ ______ ¢ ______ ¢

2. ______ ¢ ______ ¢

3. ______ ¢ ______ ¢ ______ ¢ ______ ¢ ______ ¢ ______ ¢

4. ______ ¢ ______ ¢ ______ ¢

5. ______ ¢ ______ ¢ ______ ¢ ______ ¢ ______ ¢

Name ______________________________

Going Shopping

Count the change to find how much money you have. Then circle the item that you can afford to buy.

= 14 ¢

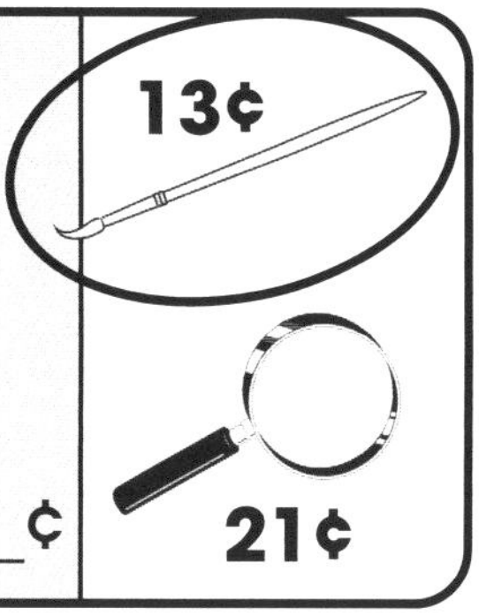

1.

= ________ ¢

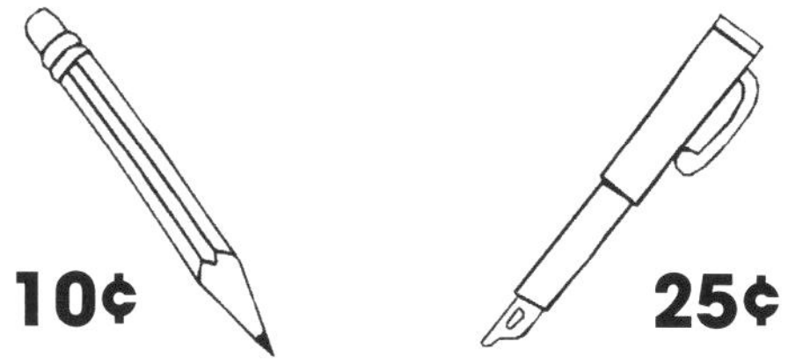

2.

= ________ ¢

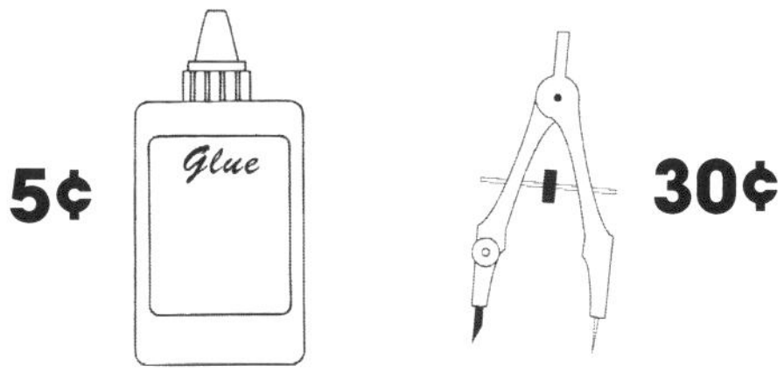

3.

= ________ ¢

4.

= ________ ¢

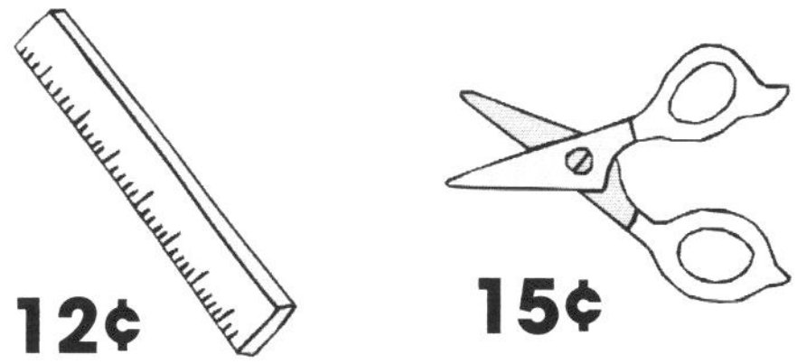

5.

= ________ ¢

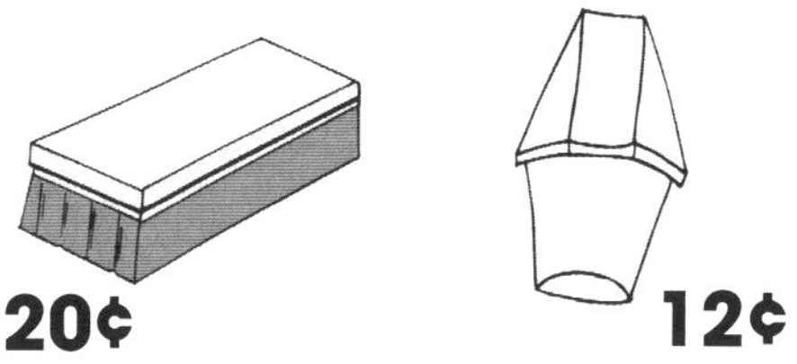

Name ________________________________

Pennies & Dimes

penny = 1¢

dime = 10¢

= 10¢

= 10¢

Count the coin values. Write the total amount on each line.

1.  = __________¢

2.  = __________¢

3.  = __________¢

4.  = __________¢

Name ______________________________

Counting By Tens

Count by tens to tell how much each group of dimes is worth. The first one is started for you.

1. 10 ¢ 20 ¢ ______ ¢

2. ______ ¢ ______ ¢ ______ ¢ ______ ¢ ______ ¢

3. ______ ¢ ______ ¢ ______ ¢ ______ ¢ ______ ¢ ______ ¢

4. ______ ¢ ______ ¢ ______ ¢ ______ ¢

5. ______ ¢ ______ ¢

Name ______________________________

Tell How Much

Count the coin values in each box.
Then write the total amount on the line.

= 13 ¢

1.

= ______ ¢

2.

= ______ ¢

3.

= ______ ¢

4.

= ______ ¢

5.

= ______ ¢

6.

= ______ ¢

7.

= ______ ¢

8.
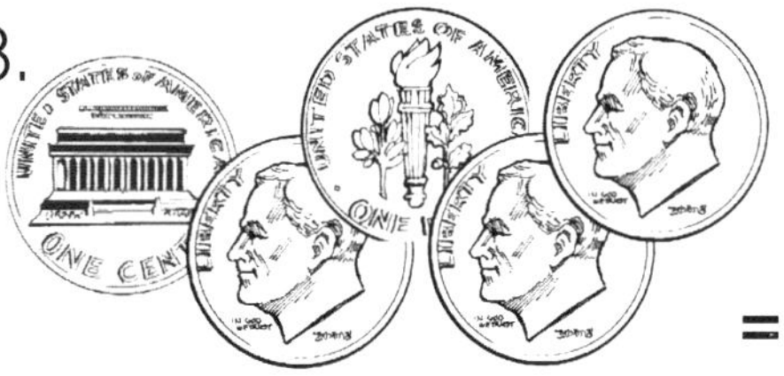
= ______ ¢

9.

= ______ ¢

10.

= ______ ¢

Name ________________________________

Going Shopping

Add the coin values to find how much each item costs. Write the total amount on the line.

10 ¢ 15 ¢

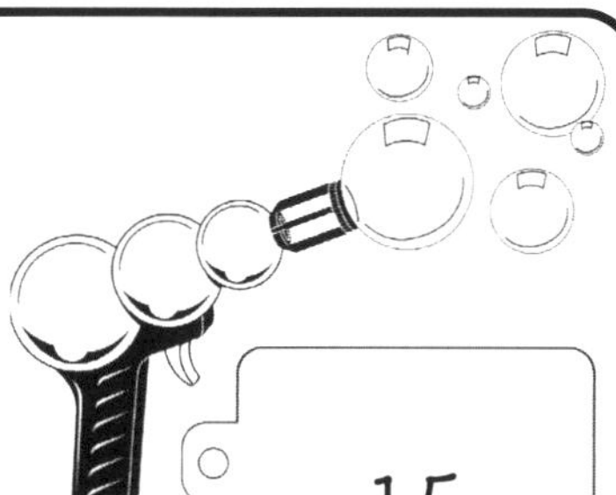

= 15 ¢

1. 

_____ ¢ _____ ¢ _____ ¢ _____ ¢

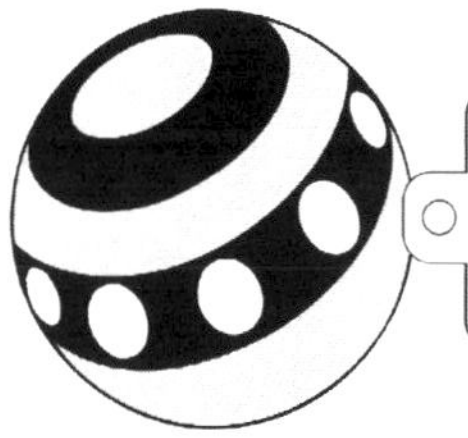

= _____ ¢

2. 

_____ ¢ _____ ¢ _____ ¢ _____ ¢

= _____ ¢

3. 

_____ ¢ _____ ¢ _____ ¢ _____ ¢

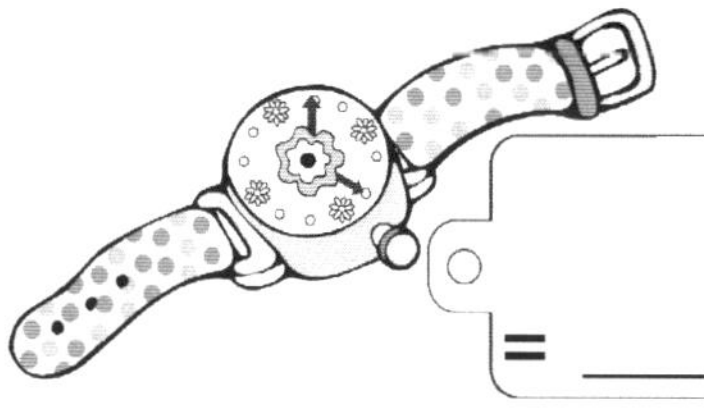

= _____ ¢

4. 

_____ ¢ _____ ¢ _____ ¢ _____ ¢

= _____ ¢

5. 

_____ ¢ _____ ¢ _____ ¢ _____ ¢

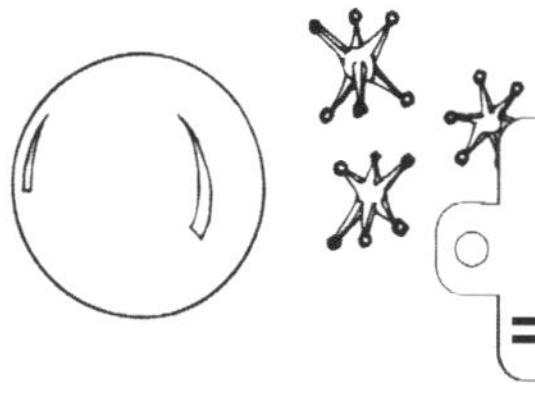

= _____ ¢

Name ______________________________

Count The Change

quarter = 25¢

 **= 25¢**

 **= 25¢**

Count the coin values. Write the total amount on each line.

1.  = __________¢

2.  = __________¢

3.  = __________¢

4.  = __________¢

Name ______________________________

Spending Money

Put an X through the coins that are needed to buy each item.

36¢

1.

35¢

5.

11¢

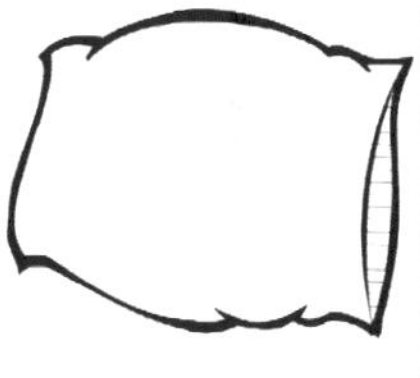

2.

32¢

6.

15¢

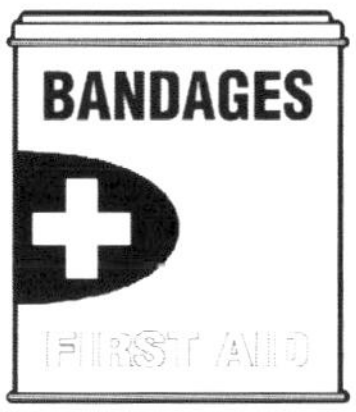

3.

45¢

7.

26¢

4.

6¢

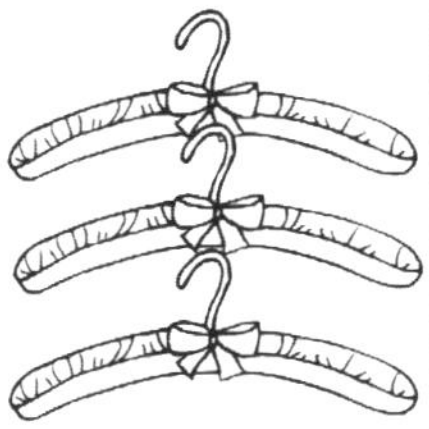

8.

51¢

Name ____________________

Coin Values

Count the coin values in each box and write the amount. Draw a line between the boxes that have matching values. The first one is done for you.

1.

= 26 ¢

2.

= ______ ¢

3.

= ______ ¢

4.

= ______ ¢

5.

= ______ ¢

6.

= ______ ¢

7.
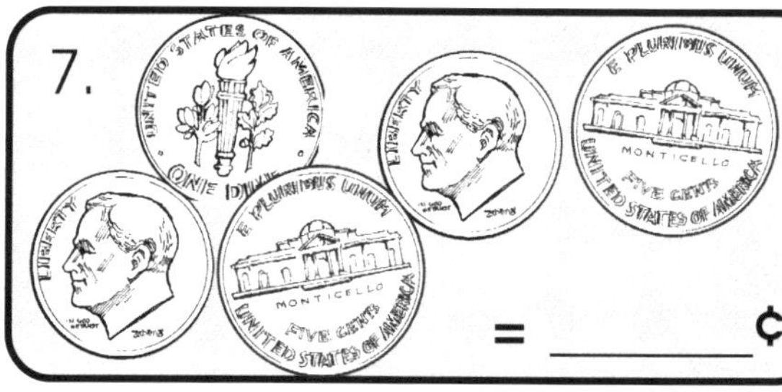

= ______ ¢

8.

= 26 ¢

Use your answers from above to complete the questions. Shade in the correct answers.

9. Which boxes have the fewest coins? ① ② ③ ④ ⑤ ⑥ ⑦ ⑧

10. Which box has the most coins? ① ② ③ ④ ⑤ ⑥ ⑦ ⑧

11. Which boxes have the highest value? ① ② ③ ④ ⑤ ⑥ ⑦ ⑧

12. Which boxes have the lowest value? ① ② ③ ④ ⑤ ⑥ ⑦ ⑧

Name ______________________________

Counting Coins

Add the coin values as you count. The first one is done for you.

1. 25 ¢ 30 ¢ 31 ¢

2. ______ ¢ ______ ¢ ______ ¢ ______ ¢ ______ ¢

3. ______ ¢ ______ ¢ ______ ¢ ______ ¢

4. ______ ¢ ______ ¢ ______ ¢ ______ ¢ ______ ¢

5. ______ ¢ ______ ¢ ______ ¢ ______ ¢

6. ______ ¢ ______ ¢ ______ ¢ ______ ¢ ______ ¢

Name ______________________________

Same Value

Put an X through the group of coins that do not have the same value.

1.   

---

2.   

---

3.   

Count the coins to complete the sentences.

4. 	One nickel is the same as __________ pennies.
5. 	One quarter is the same as __________ nickels.
6. 	Two nickels are the same as __________ dime.
7. 	__________ dimes and __________ nickel are the same as one quarter.

Name ______________________________

Half-Dollars

half-dollar = 50¢

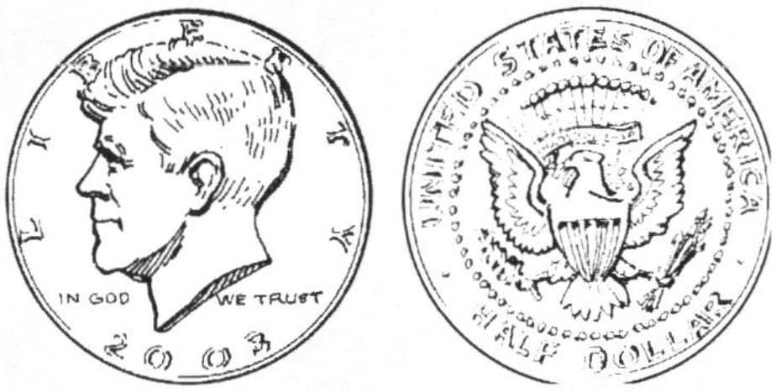

= **50¢**

= **50¢**

= **50¢**

Count the coin values in each box. Put an X through the boxes that equal 50¢.

1.

= ________¢

4.

= ________¢

2.

= ________¢

5.

= ________¢

3.

= ________¢

6.

= ________¢

Name ______________________

Adding With Coins

Write the value of each coin on the line below it. Then add the coin values to find the total.

+ + =

25 + 10 + 1 = 36 ¢

1. 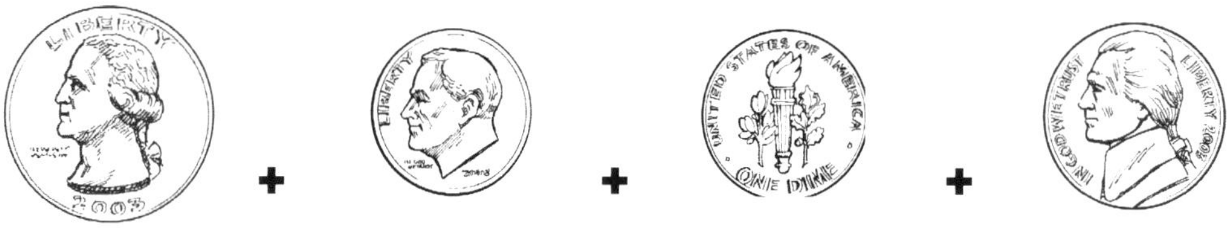

____ + ____ + ____ + ____ = ____ ¢

2. 

____ + ____ + ____ + ____ = ____ ¢

3. 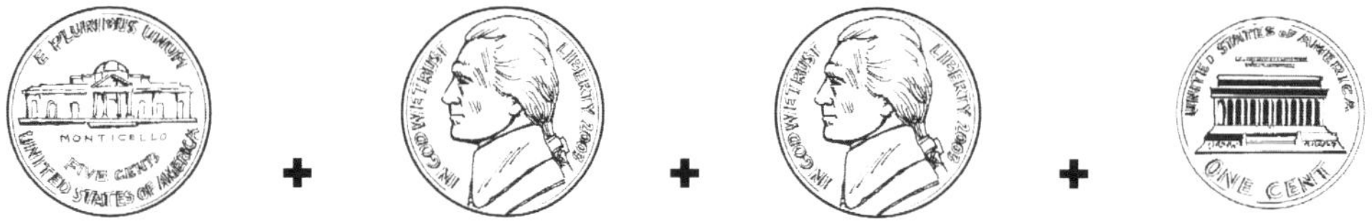

____ + ____ + ____ + ____ = ____ ¢

4. 

____ + ____ + ____ + ____ = ____ ¢

Name ______________________________

Spending Money

Put an X through the coins that are needed to buy each item.

	**12¢** 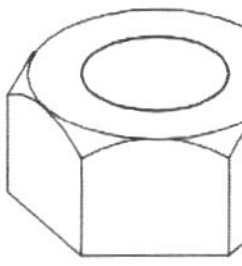		
1. 	**40¢** 	5. 	**21¢** 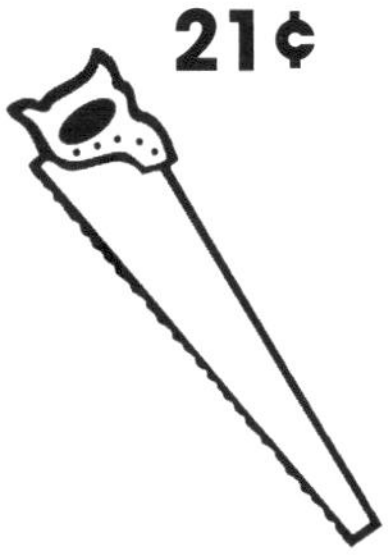
2. 	**27¢** 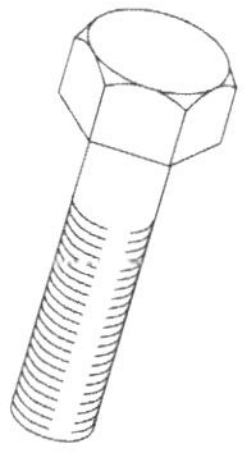	6. 	**30¢** 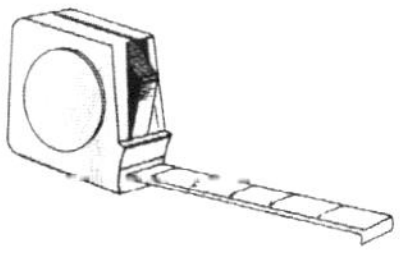
3. 	**51¢** 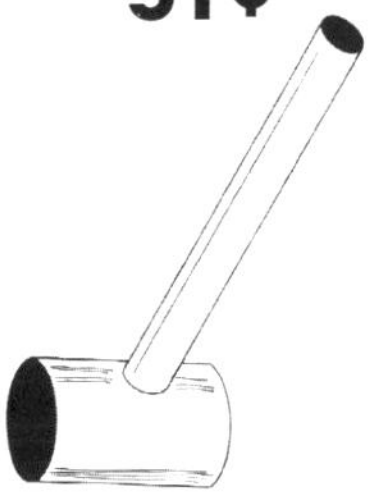	7. 	**46¢** 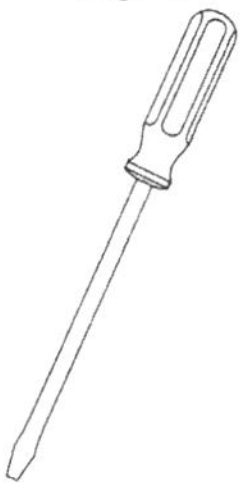
4. 	**31¢** 	8. 	**61¢** 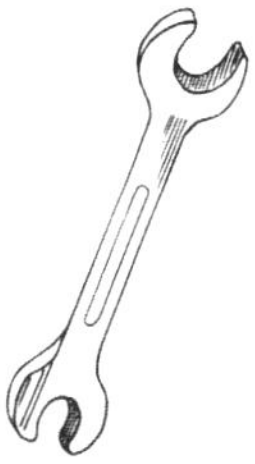

Name ______________________________

Making Connections

Draw a line and match the coins to the correct coin values. The first one is done for you.

1. 

2. 

3. 

4. 

5. 

6. 

7. 

8. 

9. 

49¢
77¢
16¢
22¢
90¢
61¢
13¢
54¢
80¢

Name ______________________________

Color The Coins

Color in the coins that equal the same amount as the coins on the left.

1.

2.

3.

4.

5.

Name ______________________

Beginning Word Problems

Use the chart to answer the questions.

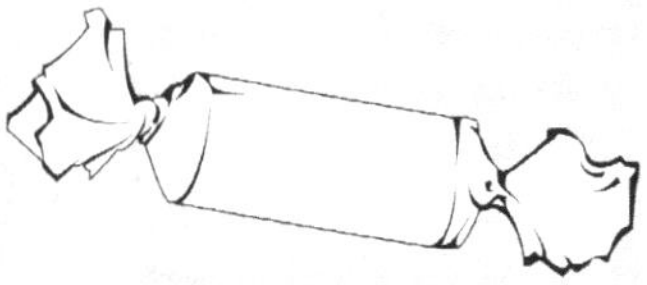 **Candy 5¢**	 **Ice Cream 50¢**	 **Pop 25¢**	 **Cookie 15¢**
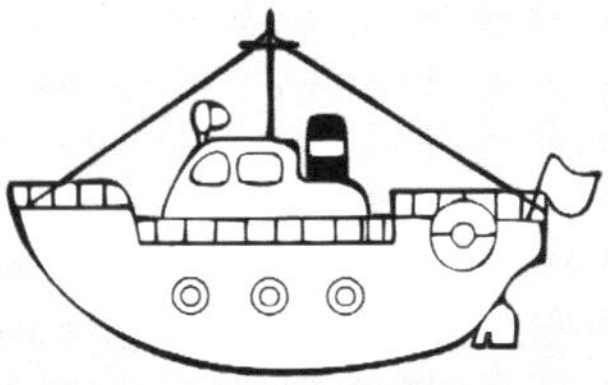 **Boat 13¢**	 **Snake 25¢**	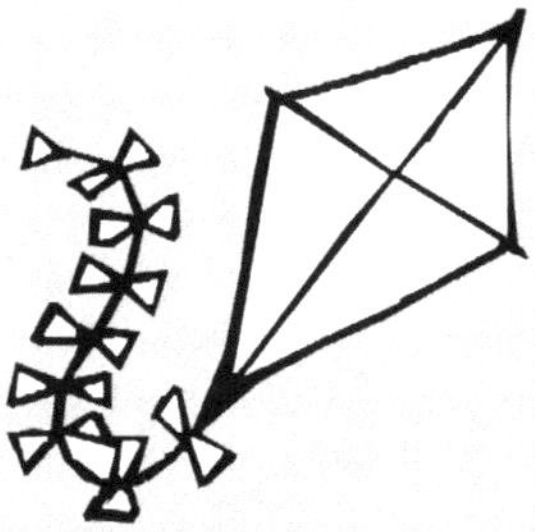 **Kite 65¢**	 **Teddy Bear 35¢**

1. Which food item costs the most?

   ____

2. Which toy costs the most?

   ____

3. Which item costs the least?

   ____

4. Which two food items add up to 55¢?

   ____

   ____

5. How much more does the kite cost than the snake?

   ____

6. How much do the pop and teddy bear cost together?

   ____

7. How much more does the pop cost than the candy?

   ____

8. How much do the boat, teddy bear, and snake cost?

   ____

Name ____________________

One Dollar

one dollar bill = $1.00

= $1.00

= $1.00

= $1.00

Count the coin values in each box. Put an X through the boxes that equal $1.00 or 100¢.

1.

= ________¢

2.

= ________¢

3.

= ________¢

4.

= ________¢

5.

= ________¢

6.

= ________¢

Name ______________________________

Color The Coins

Color in the coins that equal the same amount as the coin on the left. Use at least one of each coin.

1.

2.

3.

Write the name of a coin or bill that is worth the same as each amount.

4. 2 nickels ________________

5. 5 dimes ________________

6. 4 quarters ________________

7. 5 pennies ________________

8. 10 pennies ________________

9. 10 dimes ________________

Name ______________________________

Tell How Much

Count the change in each box. Then write the total amount on the line.

= 96 ¢

1. 

= ________¢

2. 

= ________¢

3. 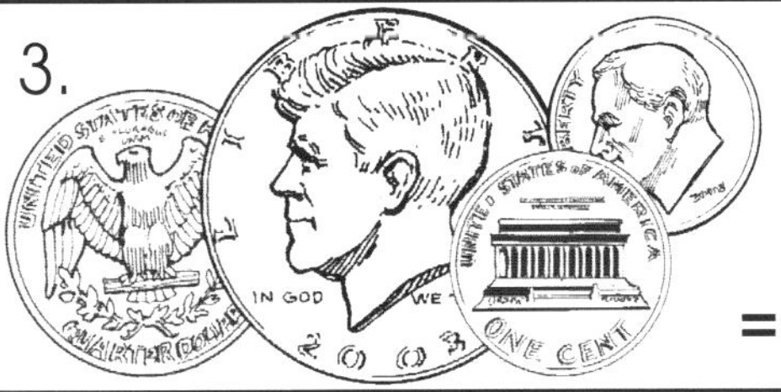

= ________¢

4. 

= ________¢

5. 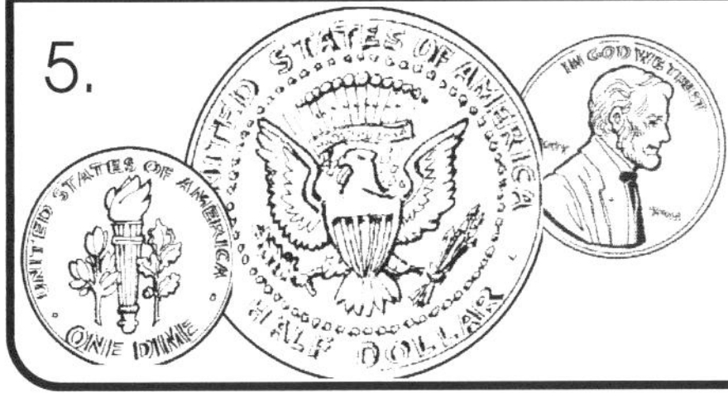

= ________¢

6. 

= ________¢

7. 

= ________¢

8. 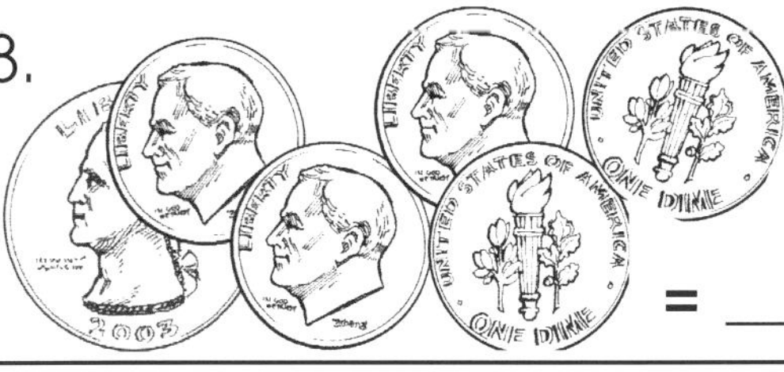

= ________¢

9. 

= ________¢

10. 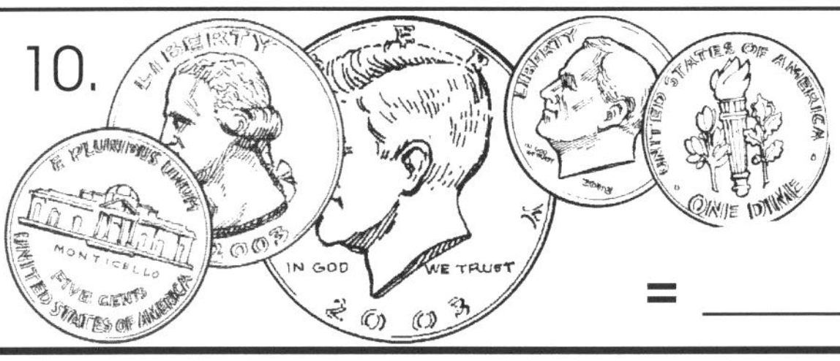

= ________¢

Name ______________________________

Going Shopping

Add the coin values to find how much each item costs. Write the total amount on the line.

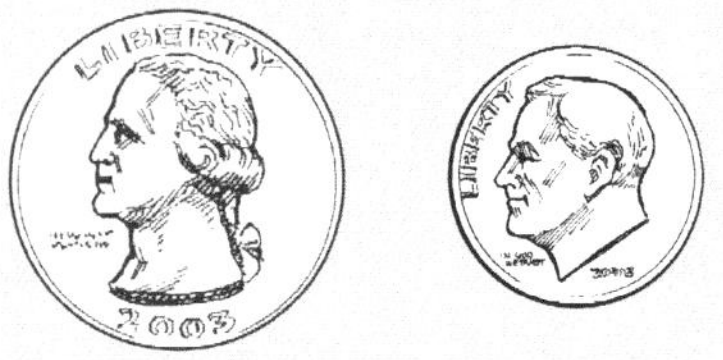

25 ¢ 35 ¢

= 35 ¢

1.

_____ ¢ _____ ¢ _____ ¢ _____ ¢

= _____ ¢

2.

_____ ¢ _____ ¢ _____ ¢ _____ ¢

= _____ ¢

3.

_____ ¢ _____ ¢ _____ ¢ _____ ¢

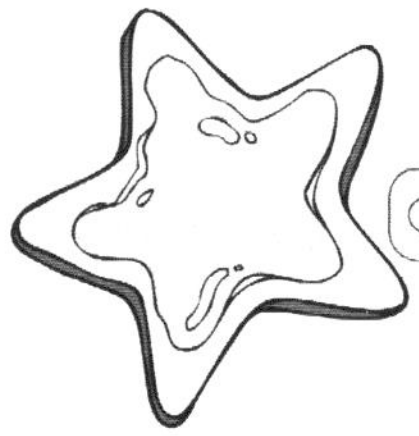

= _____ ¢

4.

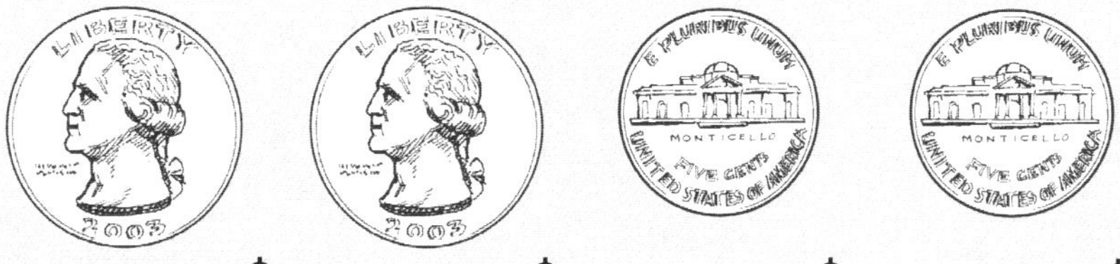

_____ ¢ _____ ¢ _____ ¢ _____ ¢

= _____ ¢

5.

_____ ¢ _____ ¢ _____ ¢ _____ ¢

= _____ ¢

Name ______________________________

Counting Money

Count the money in each box. Then circle the correct answer.

95¢

77¢ (circled)

75¢

1. 

49¢ **55¢** **41¢**

2. 

65¢ **75¢** **50¢**

3. 

$1.00 **95¢** **85¢**

4. 

76¢ **88¢** **32¢**

5. 

52¢ **50¢** **49¢**

6. 

50¢ **45¢** **65¢**

7. 

90¢ **95¢** **85¢**

8. 

60¢ **61¢** **55¢**

Name ______________________________

Greater Than/Less Than

Draw a circle around the larger value. The first one is done for you.

1. two nickels **or** three dimes
2. five pennies **or** two quarters
3. five nickels **or** two dimes
4. three quarters **or** two half-dollars
5. one dollar **or** two quarters
6. five dimes **or** five pennies

Draw a circle around the smaller value.

7. ten pennies **or** 

8. twenty nickels **or** 

9. ten dimes **or** 

10. two quarters **or** 

Name ______________________________

Largest Value

Draw a circle around the larger value. The first one is done for you.

1. 59¢ **or** five dimes

2. 25¢ **or** three quarters

3. 40¢ **or** five dimes

4. 19¢ **or** two dimes

5. 99¢ **or** one dollar

6. 5¢ **or** four pennies

7. 14¢ **or** three dimes

8. 35¢ **or** one quarter

9. 88¢ **or** two half-dollars

10. 79¢ **or** three dimes

---

11. 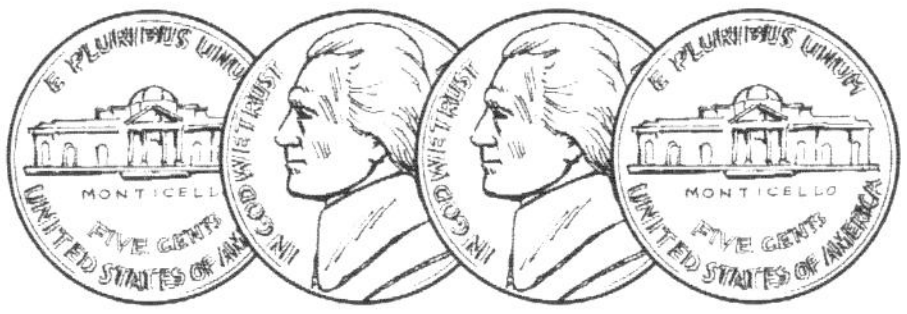 **or** 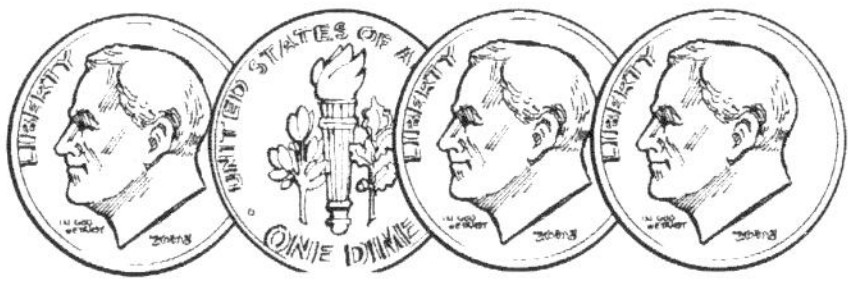

12.  **or** 

13.  **or** 

Name ______________________________

Add 'Em Up!

Write the coin values next to each coin. Then add the values to get the total. Write the amount in the shaded box. The first one is done for you.

1.

 = 25 ¢

 = 5 ¢

 = 1 ¢

\+

31 ¢

2.

 = ¢

 = ¢

 = ¢

\+

¢

3.

 = ¢

 = ¢

 = ¢

\+

¢

4.

 = ¢

 = ¢

 = ¢

\+

¢

5.

 = ¢

 = ¢

 = ¢

\+

¢

6.

 = ¢

 = ¢

 = ¢

\+

¢

Name ______________________________

Add 'Em Up!

Write the coin values next to each coin. Then add the values to get the total. Write the amount in the shaded box.

1.

 = ___ ¢

 = ___ ¢

 = ___ ¢

 = ___ ¢

\+

___ ¢

2.

 = ___ ¢

 = ___ ¢

\+

___ ¢

3.

 = ___ ¢

 = ___ ¢

 = ___ ¢

 = ___ ¢

\+

___ ¢

4.

 = ___ ¢

 = ___ ¢

\+

___ ¢

5.

 = ___ ¢

 = ___ ¢

 = ___ ¢

 = ___ ¢

 = ___ ¢

\+

___ ¢

Name ______________________________

Write It Right

Money values can be written in cents, dollar values, and words.

Example:

 = 25¢ = \$.25 = twenty-five cents

= 100¢ = \$1.00 = one dollar

Write the values a different way.

1. \$1.00

____

2. three dollars

____

3. 29¢

____

4. \$.85

____

5. \$.02

____

6. 40¢

____

7. four dollars

____

8. \$.17

____

9. 66¢

____

10. one dollar and ten cents

____

Name ______________________________

Write It Two Ways

Write the coin values two different ways.

1. $.59

____

____

2. 32¢

____

____

3. seventeen cents

____

____

4. $1.25

____

____

5. 57¢

____

____

6. eighty cents

____

____

7. $.99

____

____

8. forty-four cents

____

____

Name ______________________________

Write It Three Ways

Count the money. Write the total amounts three different ways.

1. 

____

____

____

2. 

____

____

____

3. 

____

____

____

4. 

____

____

____

5. 

____

____

____

6. 

____

____

____

Name ______________________________

Write It Three Ways

Count the money. Write the total amounts three different ways.

1. 

2. 

3. 

4. 

5. 

6. 

Name ______________________________

It's In The Bag!

Look at the money on each bag. Circle the values that are the same.

1.
27¢
twenty-six cents
$.27

2.
$.05
five cents
50¢

3.
$.47
49¢
forty-nine cents

4.
$.13
thirty-three cents
33¢

5.
$.01
two cents
1¢

6.
$1.50
$.15
one dollar and fifty cents

7.
twenty cents
2¢
$.20

8.
$.60
sixty cents
6¢

9.
one dollar
$.10
$1.00

Name ______________________________

Spending Money

Put an X through the coins that are needed to buy each item.

12¢

1. 

75¢

2. 

80¢

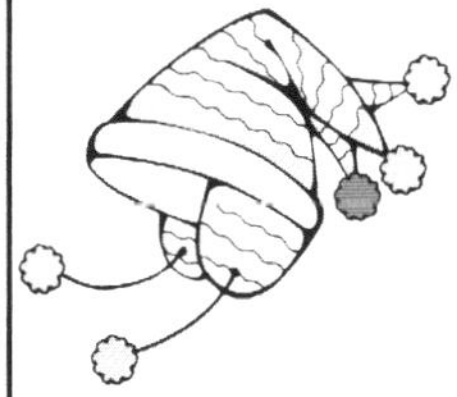

3. 

40¢

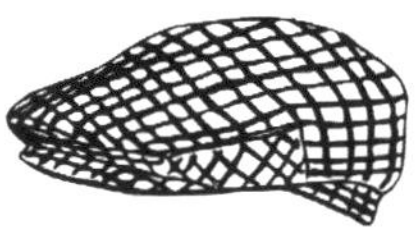

4. 

22¢

5. 

99¢

6. 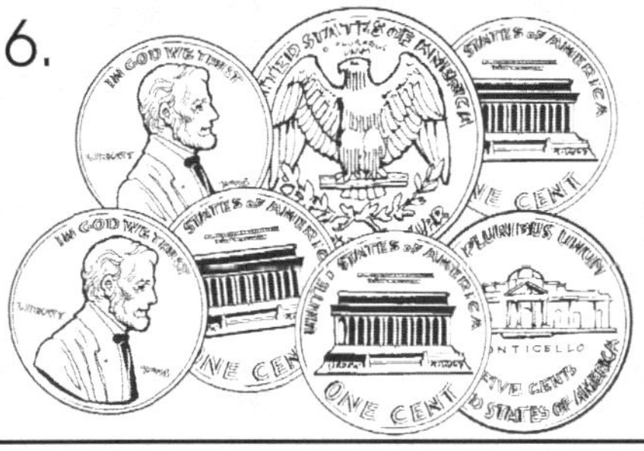

10¢

7. 

35¢

8. 

17¢

Name ______________________________

Beginning Word Problems

Use the chart to answer the questions.

Earmuffs 75¢

Coat 99¢

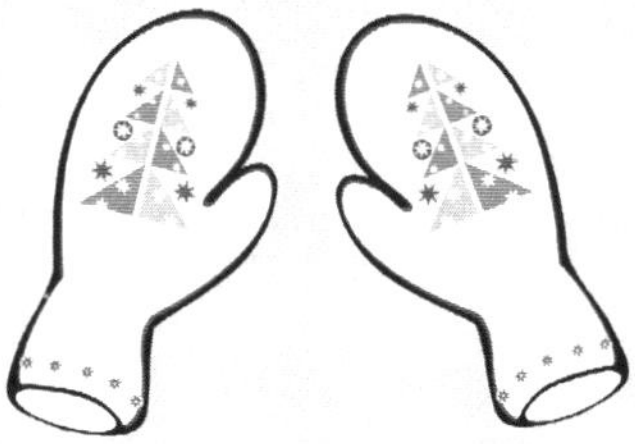
Mittens 32¢

Umbrella 18¢

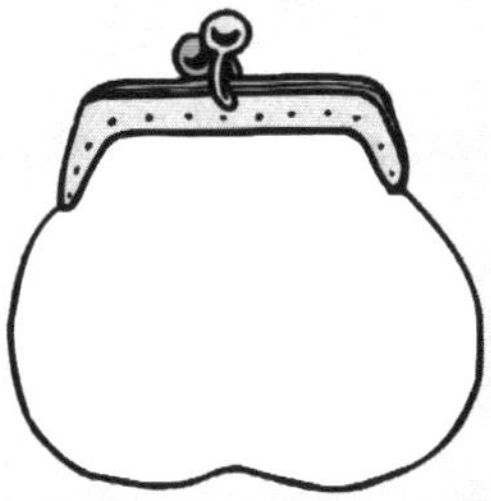
Coin Purse 19¢

Fan 27¢

Glasses 17¢

Necktie 97¢

1. What three items cost less than 20¢ a piece?

   ____

   ____

   ____

2. How much more do the earmuffs cost than the mittens?

   ____

3. What is the total cost of the coat and necktie?

   ____

4. How much more does the fan cost than the umbrella?

   ____

5. What two items can you buy with 50¢?

   ____

   ____

6. Which item costs the most?

   ____

7. Which item costs the least?

   ____

Name ________________________________

Color The Coins

Color the coins that are needed to buy each item.

20¢

1.

98¢

4.

42¢

2.

85¢

5.

76¢

3.

35¢

6.

88¢

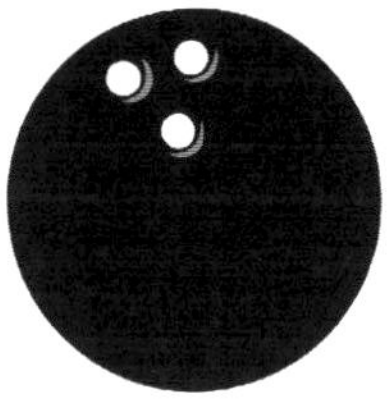

Name ______________________________

Going Shopping

Add the coin values to find how much each item costs. Write the total amount on the line.

25 ¢ 26 ¢ = 26 ¢

1. 

______ ¢ ______ ¢ ______ ¢ ______ ¢ ______ ¢

= ______ ¢

2. 

______ ¢ ______ ¢ ______ ¢ ______ ¢ ______ ¢

= ______ ¢

3. 

______ ¢ ______ ¢ ______ ¢ ______ ¢ ______ ¢

= ______ ¢

4. 

______ ¢ ______ ¢ ______ ¢ ______ ¢ ______ ¢

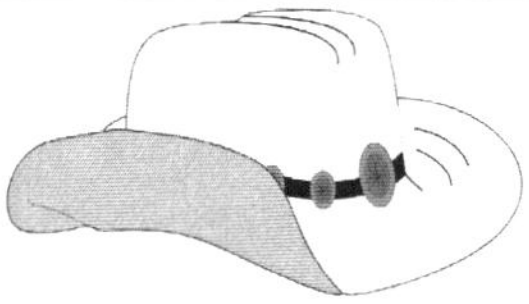

= ______ ¢

5. 

______ ¢ ______ ¢ ______ ¢ ______ ¢ ______ ¢

= ______ ¢

Name ______________________________

Going Shopping

Count the coin values to find how much money you have. Then circle the item that you can afford to buy.

= 14 ¢

20¢

13¢

1.

= ________¢

25¢

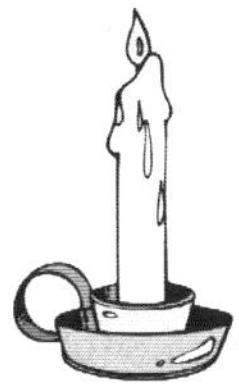

19¢

2.

= ________¢

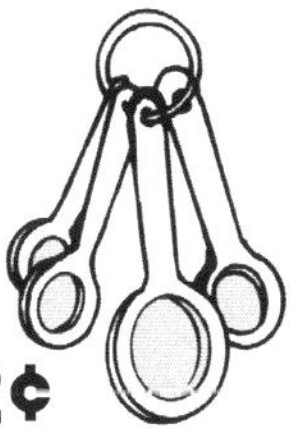

32¢

35¢

3.

= ________¢

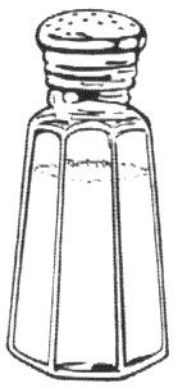

45¢

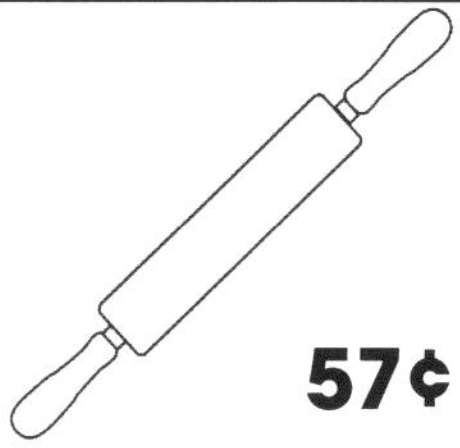

57¢

4.

= ________¢

88¢

71¢

5.

= ________¢

90¢

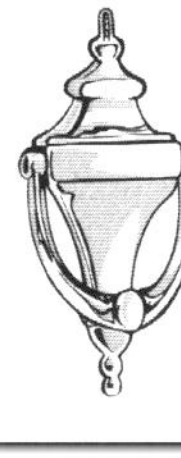

76¢

Name ______________________________

Counting Money

Count the coins in each box.
Then circle the correct amount.

92¢ (circled)

51¢

83¢

1.

$1.01

98¢

$1.06

4.

$1.05

92¢

66¢

2.

$1.50

80¢

$1.15

5.

$1.02

$1.07

85¢

3.

$1.00

81¢

55¢

6.

49¢

99¢

$1.04

Name ______________________________

Spending Money

Use the chart to answer the questions.

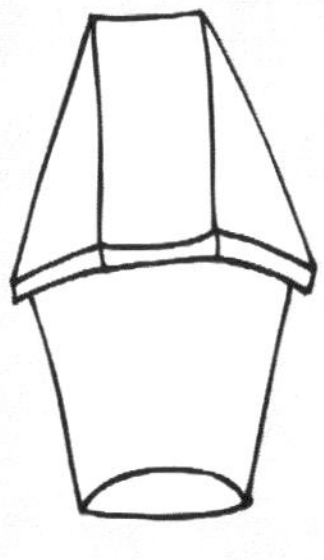

Eraser 34¢

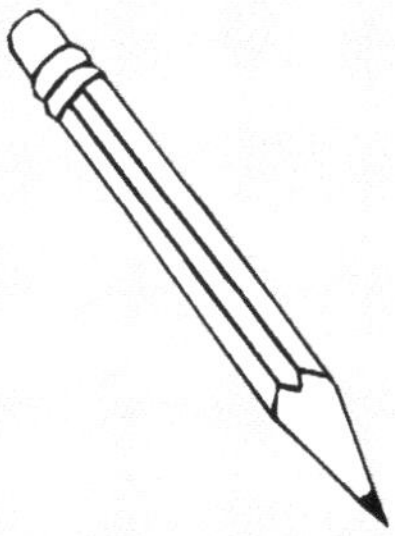

Pencil 15¢

Tea 25¢

Earrings 44¢

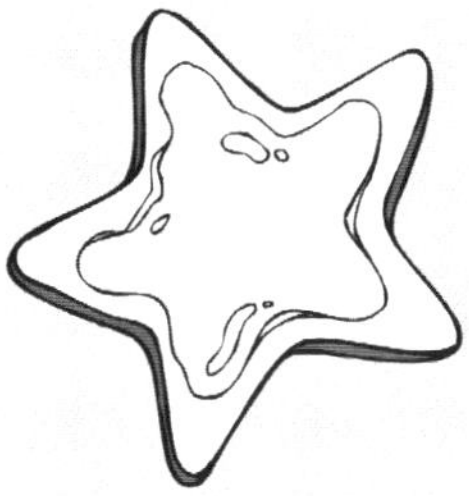

Cookie 10¢

Ring 42¢

Glue 20¢

Dinosaur 30¢

1. How much do two pencils cost?

____

2. How much do four cookies cost?

____

3. How much do three rings cost?

____

4. How much does one cookie and one cup of tea cost?

____

5. How much does one eraser and two pencils cost?

____

6. How much does one ring and one pair of earrings cost?

____

7. How much do four dinosaurs cost?

____

8. Which item costs the most?

____

Name ______________________________

Coin Values

Count the coin values in each box and write the amount. Then, draw a line between the boxes that have matching values.

1.  = ________¢

2.  = ________¢

3.  = ________¢

4.  = ________¢

5.  = ________¢

6.  = ________¢

7.  = ________¢

8.  = ________¢

Use your answers from above to complete the questions. Shade in the correct answers.

9. Which box has the fewest coins? ① ② ③ ④ ⑤ ⑥ ⑦ ⑧

10. Which box has the most coins? ① ② ③ ④ ⑤ ⑥ ⑦ ⑧

11. Which boxes have the highest value? ① ② ③ ④ ⑤ ⑥ ⑦ ⑧

12. Which boxes have the lowest value? ① ② ③ ④ ⑤ ⑥ ⑦ ⑧

Name ______________________________

Largest Value

Draw a circle around the larger value.

1.	59¢	**or**	five dimes
2.	30¢	**or**	two quarters
3.	50¢	**or**	four dimes
4.	24¢	**or**	one quarter
5.	3¢	**or**	one nickel
6.	25¢	**or**	two dimes, one nickel, and one penny
7.	45¢	**or**	four dimes and four pennies
8.	35¢	**or**	one quarter and three dimes
9.	16¢	**or**	two nickels and one penny
10.	81¢	**or**	three quarters and one nickel

---

11.  **or** 

12.  **or** 

13.  **or** 

Name ________________________________

Vending Madness

The snack machine only accepts nickels. Use the snack machine to answer the following questions.

1. How many nickels do you need to buy yogurt?

   ____

2. How many nickels do you need to buy pretzels?

   ____

3. How many nickels do you need to buy peanuts?

   ____

The jukebox only accepts dimes. Use the jukebox to answer the following questions.

4. How many dimes do you need to listen to an Elvis song?

   ____

5. How many dimes do you need to listen to an Alicia Keys song?

   ____

6. How many dimes do you need to listen to a John Mayer song?

   ____

Name ______________________________

Piggy Bank

Use the piggy bank to answer the questions.

1. How much money is in your piggy bank?

   ____

2. How many nickels are in your piggy bank?

   ____

3. How many pennies are in your piggy bank?

   ____

4. Cross out 50¢ in your piggy bank. How much money do you have left?

   ____

5. Next, cross out 87¢ in your piggy bank. How much money do you have left?

   ____

6. Then, cross out 30¢ in your piggy bank. How much money do you have left?

   ____

7. Your mom gives you $1.25 for doing your chores. Draw $1.25 in your piggy bank. Now, how much money do you have in your piggy bank?

   ____

Name ____________________________

Post-Test

1. What is this coin's value?

- (A) 1¢
- (B) 5¢
- (C) 10¢
- (D) 25¢

2. What is this coin's name?

- (A) penny
- (B) nickel
- (C) dime
- (D) quarter

3. Which coin has the largest value?

(A) 

(C) 

(B) 

(D) 

4. Which of these values is not same?

- (A) 33¢
- (B) $33
- (C) thirty-three cents
- (D) $.33

5. What is the total of one nickel and one dime?

- (A) 20¢
- (B) 10¢
- (C) 15¢
- (D) 11¢

6. Dan bought gum for 10¢ and a candy bar for 25¢.

How much did he spend?

- (A) 35¢
- (B) 45¢
- (C) 40¢
- (D) 30¢

7. Sid bought a pen for a half-dollar. Lee bought a pencil for a quarter.

How much more did Sid spend than Lee?

- (A) 50¢
- (B) 40¢
- (C) 25¢
- (D) 10¢

8. Ben has one penny, one nickel, and one dime.

How much more does he need to buy a game that costs 50¢?

- (A) 15¢
- (B) 24¢
- (C) 25¢
- (D) 34¢

PG 1 **1)** dime **2)** nickel **3)** half-dollar **4)** quarter **5)** penny **6)** quarter **7)** penny **8)** dime **9)** nickel **10)** half-dollar

PG 2 **1)** nickel **2)** pennies **3)** dimes **4)** dimes, nickel **5)** quarter **6)** quarters **7)** half-dollar **8)** pennies, nickel

PG 3 **1)** nickel **2)** quarter **3)** dime **4)** half-dollar **5)** penny

PG 4 **1)** 7¢ **2)** 9¢ **3)** 8¢ **4)** 10¢

PG 5 **1)** 6¢ **2)** 5¢ **3)** 10¢ **4)** 6¢ **5)** 3¢ **6)** 8¢ **7)** 9¢ **8)** 16¢ **9)** 13¢ **10)** 11¢

PG 6 **1)** 15¢ 20¢ **2)** 5¢ 10¢ **3)** 5¢ 10¢ 15¢ 20¢ 25¢ 30¢ **4)** 5¢ 10¢ 15¢ **5)** 5¢ 10¢ 15¢ 20¢ 25¢

PG 7 **1)** 15¢ pencil **2)** 8¢ glue **3)** 25¢ crayons **4)** 12¢ ruler **5)** 19¢ pencil eraser

PG 8 **1)** 12¢ **2)** 14¢ **3)** 13¢ **4)** 20¢

PG 9 **1)** 30¢ **2)** 10¢ 20¢ 30¢ 40¢ 50¢ **3)** 10¢ 20¢ 30¢ 40¢ 50¢ 60¢ **4)** 10¢ 20¢ 30¢ 40¢ **5)** 10¢ 20¢

PG 10 **1)** 11¢ **2)** 20¢ **3)** 10¢ **4)** 11¢ **5)** 2¢ **6)** 50¢ **7)** 23¢ **8)** 41¢ **9)** 12¢ **10)** 22¢

PG 11 **1)** 10¢ 20¢ 25¢ 26¢ **2)** 10¢ 20¢ 30¢ 31¢ **3)** 10¢ 15¢ 20¢ 21¢ **4)** 5¢ 10¢ 15¢ 20¢ **5)** 10¢ 20¢ 25¢ 30¢

PG 12 **1)** 36¢ **2)** 50¢ **3)** 31¢ **4)** 60¢

PG 13 **1)** 3 dimes, 1 nickel **2)** 1 quarter, 1 nickel, 2 pennies **3)** 1 quarter, 2 dimes **4)** 1 nickel, 1 penny **5)** 2 nickels, 1 penny **6)** 1 dime, 1 nickel **7)** 1 quarter, 1 penny **8)** 2 quarters, 1 penny

PG 14 **2)** 31¢ **3)** 50¢ **4)** 40¢ **5)** 50¢ **6)** 31¢ **7)** 40¢ **9)** 5, 8 **10)** 7 **11)** 3, 5 **12)** 1, 8

PG 15 **2)** 25¢ 50¢ 60¢ 65¢ 70¢ **3)** 25¢ 35¢ 45¢ 46¢ **4)** 25¢ 50¢ 75¢ 76¢ 77¢ **5)** 10¢ 20¢ 25¢ 30¢ **6)** 25¢ 50¢ 55¢ 56¢ 57¢

PG 16 **1)** cross out 35¢ **2)** cross out 55¢ **3)** cross out 36¢ **4)** 5 **5)** 5 **6)** 1 **7)** 2, 1

PG 17 **1)** 62¢ **2)** 50¢ - cross out **3)** 50¢ - cross out **4)** 80¢ **5)** 46¢ **6)** 75¢

PG 18 **1)** 25 + 10 + 10 + 5 = 50¢
2) 25 + 25 + 5 + 1 = 56¢
3) 5 + 5 + 5 + 1 = 16¢
4) 10 + 10 + 10 + 1 = 31¢

PG 19 **1)** 1 quarter, 1 dime, 1 nickel **2)** 1 quarter, 2 pennies **3)** 2 quarters, 1 penny **4)** 3 dimes, 1 penny **5)** 1 dimes, 2 nickels, 1 penny **6)** 1 quarter, 1 nickel **7)** 1 quarter, 2 dimes, 1 penny **8)** 2 quarters, 1 dime, 1 penny

PG 20 **2)** 13¢ **3)** 80¢ **4)** 16¢ **5)** 90¢ **6)** 22¢ **7)** 77¢ **8)** 49¢ **9)** 61¢

PG 21 **1)** 5 pennies **2)** 2 nickels **3)** 10 pennies **4)** 5 nickels **5)** 5 dimes

PG 22 **1)** ice cream **2)** kite **3)** candy **4)** ice cream and candy **5)** 40¢ **6)** 60¢ **7)** 20¢ **8)** 73¢

PG 23 **1)** 56¢ **2)** 50¢ **3)** 41¢ **4)** 90¢ **5)** 96¢ **6)** 100¢ - cross out

PG 24 **1)** 1 nickel, 5 pennies **2)** 1 dime, 2 nickels, 5 pennies **3)** 1 quarter, 2 dimes, 1 nickel **4)** dime **5)** half-dollar **6)** one dollar **7)** nickel **8)** dime **9)** one dollar

PG 25 **1)** 56¢ **2)** 90¢ **3)** 86¢ **4)** 51¢ **5)** 61¢ **6)** 57¢ **7)** 65¢ **8)** 75¢ **9)** 31¢ **10)** 100¢

PG 26 **1)** 25¢ 35¢ 45¢ 50¢ **2)** 25¢ 50¢ 55¢ 56¢ **3)** 10¢ 20¢ 30¢ 31¢ **4)** 25¢ 50¢ 55¢ 60¢ **5)** 25¢ 50¢ 75¢ 76¢

PG 27 **1)** 41¢ **2)** 65¢ **3)** 85¢ **4)** 76¢ **5)** 52¢ **6)** 45¢ **7)** 95¢ **8)** 60¢

PG 28 **2)** two quarters **3)** five nickels **4)** two half-dollars **5)** one dollar **6)** five dimes **7)** ten pennies **8)** 3 quarters, 1 dime **9)** 1 half-dollar, 2 dimes, 1 penny **10)** two quarters

PG 29 **2)** three quarters **3)** five dimes **4)** two dimes **5)** one dollar **6)** 5¢ **7)** three dimes **8)** 35¢ **9)** two half-dollars **10)** 79¢ **11)** 4 dimes **12)** 1 half-dollar, 1 quarter **13)** 2 quarters, 1 dimes, 1 penny

PG 30 **2)** 10 + 5 + 5 = 20¢
3) 25 + 10 + 10 = 45¢
4) 25 + 25 + 1 = 51¢
5) 5 + 5 + 5 = 15¢
6) 25 + 1 + 1 = 27¢

PG 31 **1)** 25 + 25 + 10 = 85¢
2) 5 + 5 = 10¢
3) 25 + 10 + 10 + 1 = 46¢
4) 25 + 1 = 26¢
5) 10 + 10 + 10 + 5 + 5 = 40¢

ANSWER KEY

PG 32 **1)** 100¢ or one dollar **2)** $3.00 or 300¢ **3)** $.29 or twenty-nine cents **4)** 85¢ or eighty-five cents **5)** 2¢ or two cents **6)** $.40 or forty cents **7)** $4.00 or 400¢ **8)** 17¢ or seventeen cents **9)** $.66 or sixty-six cents **10)** $1.10 or 110¢

PG 33 **1)** 59¢, fifty-nine cents **2)** $.32, thirty-two cents **3)** 17¢, $.17 **4)** 125¢, one dollar and twenty-five cents **5)** $.57, fifty-seven cents **6)** 80¢, $.80 **7)** 99¢, ninety-nine cents **8)** 44¢, $.44

PG 34 **1)** 7¢, $.07, seven cents **2)** 1¢, $.01, one cent **3)** 101¢, $1.01, one dollar and one cent **4)** 30¢, $.30, thirty cents **5)** 41¢, $.41, forty-one cents **6)** 45¢, $.45, forty-five cents

PG 35 **1)** 100¢, $1.00, one dollar **2)** 63¢, $.63, sixty-three cents **3)** 79¢, $.79, seventy-nine cents **4)** 50¢, $.50, fifty cents **5)** 95¢, $.95, ninety-five cents **6)** 106¢, $1.06, one dollar and six cents

PG 36 **1)** 27¢, $.27 **2)** $.05, five cents **3)** 49¢, forty-nine cents **4)** thirty-three cents, 33¢ **5)** $.01, 1¢ **6)** $1.50, one dollar and fifty cents **7)** twenty cents, $.20 **8)** $.60, sixty cents **9)** one dollar, $1.00

PG 37 **1)** 2 quarters, 2 dimes, 1 nickel **2)** 1 half-dollar, 1 quarter, 1 nickel **3)** 4 dimes **4)** 2 dimes, 2 pennies **5)** 1 half-dollar, 1 quarter, 2 dimes, 4 pennies **6)** 1 nickel, 5 pennies **7)** 2 dimes, 3 nickels **8)** 1 dime, 1 nickel, 2 pennies

PG 38 **1)** coin purse, glasses, umbrella **2)** 43¢ **3)** $1.96 **4)** 9¢ **5)** *Answers will vary.* **6)** coat **7)** glasses

PG 39 **1)** 3 quarters, 2 dimes, 3 pennies **2)** 1 half-dollar, 1 quarter, 2 nickels **3)** 1 quarter, 2 nickels **4)** 1 quarter, 1 dime, 1 nickel, 2 pennies **5)** 1 quarter, 5 dimes, 1 penny **6)** 3 quarters, 1 dime, 3 pennies

PG 40 **1)** 25¢ 50¢ 75¢ 85¢ 95¢ **2)** 25¢ 50¢ 60¢ 65¢ 66¢ **3)** 25¢ 26¢ 27¢ 28¢ 29¢ **4)** 25¢ 50¢ 75¢ 100¢ 105¢ **5)** 25¢ 50¢ 75¢ 80¢ 81¢

PG 41 **1)** 22¢ candle **2)** 33¢ measuring cups **3)** 46¢ salt shaker **4)** 71¢ iron **5)** 79¢ door knocker

PG 42 **1)** $1.01 **2)** 80¢ **3)** 55¢ **4)** 66¢ **5)** $1.07 **6)** 49¢

PG 43 **1)** 30¢ **2)** 40¢ **3)** 126¢ **4)** 35¢ **5)** 64¢ **6)** 86¢ **7)** 120¢ **8)** earrings

PG 44 **1)** 26¢ **2)** 55¢ **3)** 21¢ **4)** 65¢ **5)** 55¢ **6)** 65¢ **7)** 26¢ **8)** 21¢ **9)** 7 **10)** 5 **11)** 4, 6 **12)** 3, 8

PG 45 **1)** 59¢ **2)** two quarters **3)** 50¢ **4)** one quarter **5)** one nickel **6)** two dimes, one nickel, and one penny **7)** 45¢ **8)** one quarter and three dimes **9)** 16¢ **10)** 81¢ **11)** 3 nickels and 1 penny **12)** 1 quarter and two dimes **13)** 1 quarter, three dimes, and one penny

PG 46 **1)** 19 nickels **2)** 15 nickels **3)** 10 nickels **4)** 9 dimes **5)** 5 dimes **6)** 11 dimes

PG 47 **1)** $4.28 **2)** 2 nickels **3)** 3 pennies **4)** $3.78 **5)** $2.91 **6)** $2.61 **7)** $3.86

Post-Test **1)** C **2)** B **3)** A **4)** B **5)** C **6)** A **7)** C **8)** D